KNOW
YOUR
CLASSIC
TRACTORS

CHRIS LOCKWOOD

FORD 7810

D0880590

Know Your Classic Tractors

First published 2013. This edition published 2021.
Old Pond Publishing is an imprint of Fox Chapel Publishers International Ltd.

Designer: Emily Kingston

ISBN 978-1-913618-14-8

A catalogue record for this book is available from the British Library.

Fox Chapel Publishing
903 Square Street
Mount Joy, PA 17552

Fox Chapel Publishers International Ltd.
3 The Bridle Way, Selsey, Chichester
West Sussex PO20 0RS, U.K.

www.oldpond.com

Printed and bound in U.S.A

Cover photo: Ford's lightweight six-cylinder 7810 was an extremely versatile tractor quickly becoming a bestseller and arguably one of the most iconic tractors of the late eighties. This one is seen ploughing wheat stubble.

Contents

Foreword

There are many conflicting views on what exactly makes a 'classic tractor'. This book predominantly covers tractors produced from the late 1960s to the late 1980s – the period most commonly thought of as the classic era.

During that time, tractors experienced huge increases in power and saw a rise in the popularity of four-wheel drive. Driver comfort – previously often overlooked – became an important consideration in the late 1980s as quiet cabs with flat floors became the industry standard.

Many of the tractors included in this book are extremely popular – legends in their own right. However, some lesser-known manufacturers' examples are also included for added interest and to give a complete overview of the machines on offer at the time.

Some of the models in this book were produced at a number of factories for different markets, but the locations provided here only cover the manufacturing location for tractors sold on the UK market.

Many of these classics can still be found hard at work on British farms, often 40 years since they left the factory. Ever-increasing interest from enthusiasts and farmers alike is helping to ensure that tractors from the classic period are preserved for many more years to come.

Chris Lockwood, Suffolk, 2021

(opposite) MB-Trac tractors produced by Mercedes-Benz from the early seventies through to the early nineties have become extremely sought after classics. This MB-Trac 1100 is seen spreading fertiliser onto winter wheat.

1

Belarus MTZ-82

Assembled and/or manufactured in:
USSR
Current situation:
Minsk Tractor Works
is still producing
Belarus tractors.

The two-wheel drive MTZ-80 and four-wheel drive MTZ-82 were first produced by the Minsk Tractor Works in 1974. At that time, all tractors produced in the USSR were exported under the 'Belarus' name.

The tractors joined the existing Belarus models in the UK in 1977 with importer UMO Belarus opting for a cream and red colour scheme to fit in with the rest of the range.

Under the bonnet a 4.75-litre four-cylinder engine developed 90 hp. This 1978 MTZ-82 is working with a 3-metre Maschio power harrow.

A Swedish-built Fernmo Hytten safety cab was initially fitted for the UK market. (See tractor on facing page.) Later versions, renamed the 800 and 820, gained a slightly taller Junkkari cab from Finland.

2

Assembled and/or manufactured in:
USA
Current situation:
Case IH is still producing tractors.

Case 2090

The American-built Case 2090, 2290 and 2390 made their UK debut in 1978. They were to be sold alongside smaller English-built David Brown 90-series models.

The heavy tractors, designed mainly for drawbar work on the American prairies, were powered by large 8.3-litre six-cylinder engines and featured powershift transmissions.

The example seen here is a 2090. The 2090 was the smallest model in the range and was rated at 120 hp, while the 2390 – the largest model – was a 171 hp tractor.

In 1983 the range was updated to become the 94-series, with the 121 hp 2094 replacing the 2090. By now fourwheel drive was fitted as standard in the UK. The tractors were ultimately replaced in 1989 by the first Case IH Magnum models in the UK.

3

Assembled and/or
manufactured in:
UK
Current situation:
Case IH is still
producing tractors.

Case IH 1394

Well-known English manufacturer David Brown was acquired by Case owners Tenneco in 1972, and the Case 94-series was launched in 1983 to replace the David Brown 90-series. Initially offered in Case white, black and red livery, the tractors were updated and turned out in red, black and silver under the Case IH name following the takeover of International by Case in 1985.

The 1394 featured a 72 hp turbocharged 3.59-litre fourcylinder engine. It was available with the standard synchromesh transmission or the Hydra-Shift transmission which gave four on-the-move clutchless changes in 3 forward ratios and 1 reverse.

This particular 1394, equipped with flotation tyres and working with a Nodet Gougis fertiliser spreader, is a Commemorative Edition example. It was produced to celebrate over 50 years of David Brown in 1988 when the factory was finally closed.

4

**Assembled and/or
manufactured in:**
France and Germany
Current situation:
Case IH is still
producing tractors.

Case IH 956XL

International launched the 56-series in 1982, which included the mid-range 956XL. In 1986, with International's agricultural side now owned by Case parent company Tenneco, the tractors appeared under the Case IH name in a red, black and silver livery, and the following year the styling was updated.

Available in two- and four-wheel drive configuration, the legendary 956XL was powered by an International 5.87-litre six-cylinder engine which produced 95 hp. The gearbox gave 16 forward and 8 reverse gears, and the XL cab provided a comfortable environment for the driver.

Initially produced in both France and Germany, from 1989 the 956XL was only made at the German factory. This example is seen baling haylage with a McCormick round baler.

5

Caterpillar
D5B SA VHP

Assembled and/or manufactured in:
France
Current situation:
Caterpillar is no longer producing agricultural tractors.

The Caterpillar Tractor Company was founded in the USA when two crawler manufacturers merged in 1925. Caterpillar crawlers were early imports into Britain and gained particular popularity during and after the Second World War.

During the 1970s, agricultural 'Special Application' (SA) versions of the D4E, D5B and D6D were offered. The mid-range D5B was powered by a 10.5-litre turbocharged six-cylinder Caterpillar engine which produced 120 hp. It was fitted with a British-made Leverton cab and three-point linkage.

The tractors were later available as Variable Horse Power (VHP) versions, like this D5B SA VHP, which benefited from increased engine power when running in the higher four gears of the six-speed transmission. The D5B tractor was upped from 120 hp to 160 hp.

6

Assembled and/or manufactured in:
UK
Current situation:
County tractors are no longer in production.

County 1474

The County name has long been synonymous with equal-wheel, four-wheel drive tractors based on Ford models. County Commercial Cars began manufacturing crawler tractors in 1948 and produced their first four-wheel drive Ford conversions from 1954.

The 1474 made its debut in 1978 when it replaced the 1454. Initially the 1474 was based on the 149 hp turbocharged Ford 9700, but from 1979 onwards it was based on the TW-20 with an increase in power to 153 hp. This example is a 'short nose' 1474. A handful of 'long nose' versions were also later produced; these were based on the Ford TW-25.

Unfortunately County went into receivership in 1983. The company was bought by various new owners and a few more tractors were made, but by 1990 production ceased.

7

David Brown 990 Selectamatic

Assembled and/or manufactured in:
UK
Current situation:
Case IH is still producing tractors.

David Brown had been involved in tractor production for almost 30 years when the orchid white and chocolate brown 990 Selectamatic was launched to replace the red and yellow 990 Implematic in 1965.

Engine power was increased to 55 hp from a 3.16-litre four-cylinder David Brown engine. The Selectamatic hydraulic system allowed the driver to select one of four hydraulic functions to be controlled using a single lever.

This 990 Selectamatic is set up in high-clearance format for row-crop work or spraying taller crops. High-clear was offered as an option from David Brown and could be factory fitted. Alternatively, the conversion could be carried out on the farm when changing between seasonal tasks.

8

Assembled and/or manufactured in:
UK
Current situation:
Case IH is still producing tractors.

David Brown 996

David Brown launched the 995 and 996 in 1971 to slot in between the 990 and 1210.

Both tractors were powered by David Brown's own 3.59-litre four-cylinder engine, which produced 64 hp. The transmission was by now synchromesh and gave 12 forward and 4 reverse speeds.

The 996 benefited from an independent leveroperated PTO clutch. During the model's nine-year production run it was fitted with a variety of cabs, including the David Brown Q-cab fitted to this tractor which is carting sugar beet.

Case owners Tenneco acquired David Brown in 1972 and the Case name soon appeared on the tractors alongside the David Brown name. This later disappeared altogether in 1983 in favour of the Case name on its own.

9

Deutz-Fahr
D 78 07 C

**Assembled and/or
manufactured in:**
Germany
Current situation:
Deutz-Fahr is still
producing tractors.

German company Deutz can trace its history back to the 1860s and the early days of internal combustion engines. Early pioneers of diesel power, Deutz later went on to produce diesel-powered tractors, beginning in the late 1920s.

In 1970 Deutz completed the acquisition of fellow German manufacturer Fahr, and from 1981 the joint brand name Deutz-Fahr was introduced.

During the 1970s the tractor range consisted of the popular D-06 series models with power outputs from 24 to 120 hp. These were updated, and the new D-07 series arrived during the early 1980s.

The flagship D 78 07 was introduced in 1981 and was powered by an air-cooled 4.1-litre four-cylinder Deutz engine. This was initially rated at 75 hp, and was later increased to 82 hp. D 78 07 C versions, like this one, had the benefit of a higher specification cab.

10

Doe 130

Well-known farm machinery dealers Ernest Doe and Sons Ltd of Essex entered the tractor market in 1958. Developed from a local farmer's design, the novel Doe Dual Power was essentially two Fordson Power Major tractors with the front axles removed that were coupled together via a turntable to allow articulated steering with both tractors controlled from the rear tractor seat.

The tractor was successful because it was able to provide over 100 hp at a time when larger farms wanted more power. The design was soon improved to become the Doe Dual Drive, or Triple D.

When new Ford tractors arrived in 1964 the Triple D was developed into the Doe 130 which used two 65 hp Ford 5000 power units to give a combined power output of 130 hp.

As more conventional high-horsepower tractors came onto the market, production of Doe tandem tractors gradually declined. By the early '70s, it had ceased altogether.

11

Assembled and/or
manufactured in:
Germany
Current situation:
Fendt is still
producing tractors.

Fendt Favorit 612 LSA Turbomatik

Fendt produced their first tractors in Germany during the 1930s. The tractors were very popular on the continent, but in Britain the company was predominantly known for their tool carriers during the 1970s. A large range of conventional tractors was also produced, and these gained more of a foothold during the 1980s.

The high-horsepower Favorit SL range was introduced in 1976 with power outputs from 85 to 150 hp. This range was updated two years later with the introduction of Favorit LS models. The range was updated again in 1984. The Favorit 612 LSA was powered by a 135 hp six-cylinder turbocharged MWM engine.

Unusually this drove through a Turbomatik fluid clutch for a smooth uptake of power when pulling away, and the gearbox gave 20 forward and 9 reverse speeds. The tractor's linkage was controlled electronically and reverse drive was offered as an option – highly advanced for the time.

12

**Assembled and/or
manufactured in:**
Italy
Current situation:
**New Holland is still
producing tractors.**

Fiat 640

Founded in Italy in 1899, Fiat produced their first tractors in 1919. Fiat crawler tractors were sold in Britain from the 1950s, and these were particularly popular during the 1970s when the company became extremely well known for them.

The iconic 'Nastro D'Oro' range of wheeled tractors was launched in 1968 and spanned 25 to 85 hp, with later new additions eventually extending the series up to 150 hp. One of the most popular models was the 64 hp 640, introduced in 1972, which was also offered with four-wheel drive as the 640 DT.

It was powered by Fiat's own 3.46-litre four-cylinder engine, and the tractor was equipped with an 8 forward and 4 reverse gearbox. For the British market the 640 was fitted with a Lambourn cab – as seen on this example. It is also equipped with wider rear wheels and is seen spreading beans onto stubble ready to be ploughed in.

13

Ford Major 4000

Assembled and/or
manufactured in:
UK
Current situation:
New Holland is still
producing tractors.

In 1964 Ford launched a new range of tractors: the 6X, or thousand series. The line-up consisted of the Dexta 2000, Super Dexta 3000, Major 4000 and Super Major 5000 with power outputs ranging from 37 to 65 hp. The model names were carried over from old Fordson models to make the new tractors more familiar to British farmers.

The 55 hp Major 4000 replaced the famous Fordson Super Major, and was powered by a 3.29-litre threecylinder engine. The standard transmission gave 8 forward speeds and 2 reverse, but the tractor could also be specified with the clutchless change-on-themove Select-O-Speed gearbox.

The Dexta and Major names were dropped in 1968 and the power of the engines was increased with the introduction of the Ford Force thousand range.

14

Assembled and/or manufactured in:
UK
Current situation:
New Holland is still producing tractors.

Ford 7000

Ford launched the iconic 7000 in 1971, slightly later than the rest of the range. It was to become the new flagship and most powerful tractor offered by Ford at the time. The 7000 was powered by a turbocharged four-cylinder engine which produced 94 hp and packed a lot of power into a compact size.

It proved very popular with British farmers and was the first Ford offered in the UK with a factory-fitted turbocharger. Outwardly, the 7000 looked very similar to the less-powerful 5000, except that the 7000 had a taller air intake and larger tyres.

For the UK market the 7000 was fitted with Ford's deluxe safety cab as standard. This can be seen on this example, which is working with a single-leg subsoiler to pull up some spring bean stubble.

15

Assembled and/or
manufactured in:
UK
Current situation:
New Holland is still
producing tractors.

Ford 4600

In 1975 Ford launched the new 600 series running from the 37 hp 2600 through to the 97 hp 7600 with improved engines. The following year new luxury Qcabs were introduced throughout the range and these were a vast step forward in driver comfort.

The 62 hp 4600 was the largest three-cylinder tractor in the range and proved a popular model for lighter farm work.

This Q-cab equipped 4600 is baling barley straw with a New Holland 378 conventional baler.

The restyled Series 10 tractors were launched in 1981, with a new gearbox, and the 4600 was replaced by the 64 hp 4610.

16

Assembled and/or
manufactured in:
Belgium
Current situation:
New Holland is still
producing tractors.

Ford TW-25

Ford's high-horsepower TW range was introduced in 1979 and consisted of the TW-10, TW-20 and TW-30. Four years later it was updated, with the three models becoming the TW-15, TW-25 and TW-35.

The mid-range TW-25 gained 10 hp more than the TW-20 that it replaced, with 163 hp from a turbocharged 6.58-litre six-cylinder engine. It was a slightly longer tractor, like the larger TW-35, with a larger fuel tank housed under a longer bonnet.

This TW-25, fitted with a front-mounted Dowdeswell press arm, is spraying spring barley with a Berthoud trailed sprayer.

In 1985 Force II versions of the TW range appeared, fitted with Super-Q luxury cabs.

17

Assembled and/or
manufactured in:
UK
Current situation:
New Holland is still
producing tractors.

Ford 7810
Generation III

The legendary 7810 joined Ford's Series 10 Force II range in 1988 just before they were replaced the following year by the Series 10 Generation III models. The 7810 had a special place in the range. Being relatively light and compact while packing plenty of power (105 hp) from a 6.57-litre six-cylinder engine, the tractor had an excellent power to weight ratio. It proved to be extremely popular with British farmers.

This tractor is ploughing with a four-furrow Ransomes TSR 300 plough.

The 7810 III was replaced by the new 100 hp 7840 in late 1991 with the introduction of the new Series 40 tractors.

The company was later re-named New Holland following the takeover of Ford by Fiat.

18

Hürlimann H-490

Assembled and/or manufactured in:
Switzerland/Italy
Current situation:
Hürlimann
tractors are still
in production.

Swiss tractor maker Hans Hürlimann built his first tractor in 1929. In 1977 the company was bought by Italian manufacturer Same and two years later it became part of the Same-Lamborghini-Hürlimann (SLH) Group.

The H-490 was mid-range model in the company's Hseries which peaked at 200 hp and had been launched in 1979 together with change in colour from red to pale green. It was a 95 hp tractor that was powered by a four-cylinder turbocharged Hürlimann engine which was water-cooled.

Production of the series was later moved to Italy and came to end in the mid-1980s when new models appeared.

19

Assembled and/or
manufactured in:
UK
Current situation:
Case IH is still
producing tractors.

International 574

In 1970 International launched the first of their World Wide series of British-built tractors which eventually ran from 35 hp to 78 hp. One of the first models to be introduced was the 574, and it would prove to be one of the most popular International tractors of the era.

It was powered by a four-cylinder 3.92-litre International engine which produced 68 hp. Three transmission options were available: the standard synchromesh with 8 forward and 4 reverse speeds, synchromesh with Torque Amplifier – a splitter which doubled the number of speeds available, and an infinitely variable hydrostatic transmission.

For the home market the tractors were fitted with Victor safety cabs, as shown in this example.

20

Assembled and/or
manufactured in:
UK
Current situation:
Case IH is still
producing tractors.

International 584

The 84-Series 'High Performers' were launched by International in 1977. The range ran from the 45 hp 384 through to the 136 hp 1246.

The 584 was one of the mid-range models of the British-built tractors in the series. It was powered by a 62 hp four-cylinder International engine. Like the 574 before it, the tractor could be specified with the Torque Amplifier to double the number of speeds offered by the standard synchromesh gearbox.

A Victor cab like the one fitted to this 584 was standard equipment, but a flat deck Super De Luxe cab made by Sekura was also offered as an option.

A hydrostatic tractor was also offered as part of the 84-Series, this being the 80 hp Hydro 84.

21 International 885 XL

Assembled and/or
manufactured in:
UK
Current situation:
Case IH is still
producing tractors.

International updated its tractor line-up in 1981 with the introduction of the Fieldforce 85-Series. Probably the main difference was the new deluxe XL Control Centre cab – fitted to the tractor on the facing page – which had been designed with driver comfort in mind. Tractors could also still be specified with the economy L model cab.

The largest tractor in the range was the 885 XL which was available in two- and four-wheel drive configuration. Under the bonnet a 4.39-litre fourcylinder International engine provided 85 hp.

In 1985, the Agricultural division of International Harvester was bought by Tenneco, who already owned Case and David Brown. With updated styling and a new red and black livery 85-Series tractors continued to be made under the Case IH name until 1990.

22

John Deere 4020 Diesel

Assembled and/or manufactured in:
USA
Current situation:
John Deere is still producing tractors.

John Deere two-cylinder tractors were imported from America from the 1920s and were particularly popular for row-crop work. It was not until the advent of Europeanbuilt John Deere tractors, following their acquisition of German-manufacturer Lanz in 1956, that they began to make a serious impact on the UK market from the mid-1960s.

A landmark tractor, the American-built John Deere 4020 replaced the 4010 in 1963 and would prove to be an extremely popular tractor the world over. A powerful tractor, it was equipped with a 106 hp six-cylinder John Deere engine. The 4020 was offered with a powershift transmission – an option in place of the standard gearbox and one that was ahead of its time.

The 4020 was soon joined by the smaller German-made 1020, 1120 and 2020. These tractors helped John Deere gain a foothold in the UK.

23

Assembled and/or
manufactured in:
Germany
Current situation:
John Deere is still
producing tractors

John Deere 3130

John Deere launched the new 30-series in 1972. The range included models produced in Europe and America as well as the largest German-built tractor – the 3130. Initially introduced with styling similar to the previous 20-series, a new rounder bonnet like that fitted to this tractor was used from the mid '70s.

The 3130 was powered by a 5.4-litre six-cylinder John Deere engine that originally produced 89 hp but was slightly increased to 92 hp in 1976. This provided power to a 12 forward 6 reverse transmission which also featured a clutchless on-the-go Hi-Lo splitter.

A new cab was offered for the tractors: the Operator's Protection Unit (OPU) which was produced by Sekura. It improved driver comfort compared to previous offerings.

24

Assembled and/or manufactured in:
Germany
Current situation:
John Deere is still producing tractors.

John Deere 2140

In 1979 John Deere launched their European-built 40-series tractor line-up, ranging from the 50 hp 1040 through to the 97 hp 3140. The most powerful fourcylinder model in the range was the 2140. This model found particular favour with British farmers.

The tractor's four-cylinder engine was turbocharged and produced 82 hp. The Power Synchron transmission with a change-on-the-move Hi-Lo splitter was offered as an option.

Initially launched with the OPU cab, the 2140 was soon available with John Deere's iconic SG2 cab – see example on facing page. With its large curved-glass windscreen, the left side of which formed part of the door, the cab looked modern and offered excellent driver comfort.

The 2140 was replaced by the 2850 in 1986 with the introduction of the rest of the German-built 50-series models.

25

Assembled and/or
manufactured in:
USA
Current situation:
John Deere is still
producing tractors.

John Deere 4440

John Deere announced the new American-built highhorsepower40-series models in 1977. With slightly more power than the 30-series tractors they replaced, the range offered in the UK ran from the 110 hp 4040 through to the 177 hp 4640.

One of the most popular models in the line-up was the 4440. This was powered by a new 155 hp 7.64-litre six-cylinder John Deere engine which was also turbocharged. The tractor was fitted with a 16-speed Quad-Range transmission as standard, but a full powershift gearbox was also an option.

This 4440 is fitted with the Sound-Guard cab. It is also equipped with optional hydrostatic four-wheel drive, which used separate hydraulic motors to drive each front wheel. The tractor is driving a Hesston big baler.

26

Assembled and/or
manufactured in:

Italy

Current situation:

Lamborghini tractors are
still in production.

Lamborghini R 653

The first Lamborghini tractor was built in 1948 and fellow Italian manufacturer Same bought the tractor business in 1972. Lamborghini tractors appeared in the UK from the mid '70s; the range on offer had power outputs from 38 to 105 hp.

The 62 hp three-cylinder 653 was introduced in the early '80s. Like most Lamborghini tractors built up until this time it was powered by an air-cooled engine, although the company would later make use of watercooled power units.

Two-wheel drive versions like this one were designated R 653 while a four-wheel drive version, the 653 DT, and crawler version, the C 653, were also produced.

27

Assembled and/or manufactured in:
UK
Current situation:
Leyland tractors are no longer produced.

Leyland 282 Synchro

In 1969 Nuffield tractors emerged under the Leyland name following the merger of British Motor Holdings and the Leyland Motor Corporation.

Introduced in 1979, the two-wheel drive 282 Synchro and the four-wheel drive 482 Synchro were additional models in the Leyland range.

Essentially an up-rated and more powerful version of the popular 272, the 282 was powered by a turbocharged version of Leyland's 3.77-litre fourcylinder engine that produced 82 hp. It was fitted with the usual 9 forward and 3 reverse synchromesh transmission.

In 1980 the colour of Leyland tractors was changed to harvest gold with the introduction of a new range.

28

Assembled and/or
manufactured in:
UK
Current situation:
Massey Ferguson is still
producing tractors.

Massey Ferguson 135

Massey Ferguson launched the groundbreaking 'Red Giant' 100-series in 1964 when the legendary 135 replaced the highly successful 35. The 135 tractor was a handy size and became extremely popular on British farms. In fact, many can still be found at work today.

Essentially a very similar machine to the previous 35X, the 135 was powered by a three-cylinder Perkins engine that initially produced 45 hp. This output was increased to 47 hp when the range was updated in 1971.

The gearbox gave 6 forward and 2 reverse speeds, but could also be specified with the change-on-the-move Multi-Power high-low gearbox to give double the speeds. This could be changed up for more speed or changed down to reduce load when pulling through tough spots or uphill.

The 135 had an extremely long production run, finally coming to an end in 1976 when it was replaced by the 47 hp 550.

29

Massey Ferguson 590

Assembled and/or manufactured in:
UK and France

Current situation:
Massey Ferguson is still producing tractors.

Having already launched the 595 in 1974, Massey Ferguson introduced the rest of the 500-series in 1976. The new tractors replaced the outgoing 100-series models, and the 590 replaced the 185 and 188.

The tractor was powered by a 4.06-litre four-cylinder Perkins engine which provided 75 hp. The standard gearbox offered 8 forward speeds, but a 12-speed Multi-Power transmission was a popular option.

Like the rest of the range, the 590 benefited from the new Supercab with its single large left-hand door, although two-door cabs were introduced towards the end of production.

The 590 was later also available with four-wheel drive. This two-wheel drive example is spreading fertiliser with an Amazone twin-disc spreader.

30

Assembled and/or manufactured in:
UK
Current situation:
Massey Ferguson is still producing tractors.

Massey Ferguson 1250

In the early 1970s Massey Ferguson introduced the impressive four-wheel drive 1200 with a 105 hp engine, articulated steering and integral cab. The tractor was replaced by the 1250 in 1979 which, although outwardly very similar, gained a bit more power along with other improvements.

Under the bonnet a 5.8-litre Perkins engine now produced 112 hp. A Multi-Power transmission came as standard and offered 12 forward speeds and 4 reverse.

The introduction by Massey Ferguson of powerful, conventional four-wheel drive tractors at a similar time caused the popularity of the articulated 1250 to suffer, and production ceased in the early 1980s.

This particular example is fitted with an aftermarket Opico turbocharger to increase the engine's power output. It has also been fitted with wide flotation tyres.

31

Assembled and/or manufactured in:
UK
Current situation:
Massey Ferguson is still producing tractors.

Massey Ferguson 690

Launched in 1982, the Massey Ferguson 600-series replaced the previous 500-series tractors. Fitted with new luxury cabs, the middle model in the initial range was the 690 which replaced the popular 590.

Like the 590, the 690 was powered by a 4.06-litre fourcylinder Perkins engine rated at 75 hp, although this was later increased to 77 hp. Two transmissions were offered: the Synchro-12, which was standard, or Multi-Power. Both offered 12 forward and 4 reverse speeds.

This four-wheel drive example is rolling ploughed land. The 690 was replaced in 1986 by the 77 hp 3060 from a new generation of tractors with electronic controls.

32

Massey Ferguson 2680

Assembled and/or manufactured in:

France

Current situation:

Massey Ferguson is still producing tractors.

The first of Massey Ferguson's high-horsepower 2000-series arrived in 1979, and before long the Frenchbuilt range included four models with power outputs spanning 93 hp to 147 hp.

Like the rest of the range, the 2680 was powered by a 5.8-litre six-cylinder Perkins engine which, for this model, was turbocharged and rated at 126 hp. A 16-speed Speedshift gearbox was provided, with twospeed powershift.

The new Super de-luxe cab with a single left-hand door featured air-conditioning as standard. In 1985 the range was replaced by the revised 2005-series, and changes included a move over to two-door cabs.

33

Assembled and/or
manufactured in:
Germany
Current situation:
Mercedes-Benz is no
longer producing tractors.

MB-trac 1300

Mercedes-Benz had already been producing its Unimog multi-purpose vehicle for over 20 years when the first MB-trac tractor was launched in the early 1970s. The new tractor had an unusual layout, with the mid-mounted cab allowing for a rear load platform which was often used for mounting implements like sprayers.

By the late 1970s a whole range of MB-tracs was in production, and the flagship model was the 1300 which had been introduced in 1976.

Power came from a 5.68-litre six-cylinder turbocharged Mercedes Benz engine which gave out 125 hp. A 12-speed gearbox was provided, which gave a fast top speed on the road of 25 mph, and the tractors had the benefit of airbrakes and suspension.

The 1300 was updated in the late 1980s but unfortunately MB-trac production came to an end in 1991.

34

Muir-Hill 121 Series II

Assembled and/or manufactured in:
UK
Current situation:
Muir-Hill tractors are no longer in production.

Muir-Hill, well known for their range of dumper trucks and loading shovels, introduced their first tractor, the four-wheel drive 101 hp 101, in 1966. Like a lot of British-built equal-wheel four-wheel drive tractors at the time it used a lot of Ford components. In 1972 it was superseded by the 120 hp 121.

Power for the 121 was provided by a 6.22-litre sixcylinder Ford engine, and Ford gearboxes were used to provide 8 forward speeds and 2 reverse. The improved 121 Series III arrived in the late '70s with a bit more power (132 hp) and a slightly larger cab.

Production of Muir-Hill tractors ceased in 1982 when the company changed hands, although a handful of improved machines were later produced by the current owners under the Myth-Holm name.

35

Assembled and/or manufactured in:
UK
Current situation:
Roadless tractors are no longer in production.

Roadless 115

Roadless Traction was another British company well known for four-wheel drive Ford-based conversions. The company originally produced half-track conversions and crawlers, later progressing onto unequal-wheel four-wheel drive tractor conversions in the mid '50s.

In 1968 Roadless launched their first equal-wheel fourwheel drive tractor: the 115. It was powered by a 115 hp six-cylinder Ford engine and used many strengthened Ford components. With the introduction of new safety regulations a Duncan safety cab was also fitted, as on this tractor.

Roadless fell into financial difficulties in 1983 and ceased production, although a handful of tractors were subsequently produced under new ownership bearing the Jewelltrac name.

36

Assembled and/or
manufactured in:
Italy
Current situation:
Same is still
producing tractors.

Same Leopard 90 Turbo

Same can trace its tractor producing history back to the late 1920s. The Italian company was a very early pioneer of four-wheel drive. Same tractors first appeared on the UK market in the mid-1960s.

In 1982 the Leopard 90 Turbo joined the range, which went up to 160 hp, to replace the Leopard 85. An 88 hp tractor, it was powered by Same's own 4.16-litre four-cylinder turbocharged engine. This was aircooled like most Same engines – a feature for which the company was particularly well known. The transmission offered 12 forward speeds and 3 reverse.

By the mid-1980s the Leopard 90 Turbo had been replaced by the new Laser 90 Turbo.

37

Assembled and/or
manufactured in:
USA
Current situation:
Case IH is still
producing tractors.

Steiger Panther ST 325

The Steiger brothers made their first articulated fourwheel drive tractor on their farm in America during the late 1950s, and within a few years commercial production had begun. The models in the Steiger range were given big cat names like Bearcat, Cougar, Lion, Tiger, Panther and Puma. The first tractors were imported into the UK in the mid-1970s.

The Panther ST 325 was part of the Series III range introduced in 1976 and was the largest model offered in the UK. It was powered by a 14.6-litre six-cylinder turbocharged Caterpillar engine that produced 325 hp.

Steiger also produced articulated tractors for Ford which were sold in Britain during the late '70s and early '80s. In 1986 the company was bought by Case IH who continued articulated tractor production under their own name.

38

Assembled and/or
manufactured in:
UK
Current situation:
Track Marshall crawlers
are no longer
in production.

Track Marshall 100

British manufacturer Marshall launched their first Track Marshall crawler in the mid-1950s and this was followed by a number of successful models. In 1975 Marshall-Fowler was bought by British Leyland and the name on the bonnet was changed to Aveling Marshall.

Aveling Marshall replaced the very popular 90 crawler with the Perkins-powered 100 and Ford-powered 105 – essentially just updated versions with increased power. The range reverted back to the Track Marshall name in 1979 when the company changed hands again.

The 100 was fitted with a six-cylinder Perkins engine that developed 100 hp and drove through a five-speed gearbox. The tractor used a different steering system to contemporary Caterpillar and Fiat crawlers.

39

**Assembled and/or
manufactured in:**
Romania
Current situation:
Universal is no longer
producing tractors.

Universal 550

Universal tractors were produced by UTB, which stands for Braov Tractor Factory in Romanian. The first tractors were imported into the UK in 1969, and by the 1970s most of the range was based on Fiat models.

The 550 was launched in 1972 together with the smaller 445, which was also produced in a crawler version.

It was powered by a 55 hp 3.12-litre four-cylinder engine and the gearbox gave 8 forward and 2 reverse speeds. A four-wheel drive version was also offered, designated the 550 DT. Like most Universal tractors sold in the UK this 550 is fitted with a Duncan safety cab.

40

Assembled and/or
manufactured in:
Poland
Current situation:
Ursus is still
producing tractors.

Ursus C-355

Ursus tractors, produced in Poland since 1922, first appeared on the UK market in the late 1960s. The C-355 was introduced in 1969 and was an improved version of the C-350, the export version of the C-4011, which was itself a license-built Zetor 4011. This C-355, fitted with a Duncan cab, dates from 1972.

The C-355 was powered by a 3.12-litre four-cylinder engine which produced 55 hp and drove through a gearbox with 10 forward and 2 reverse speeds.

In Poland the C-355 received some minor updates in 1976 to become the C-360 but was still sold as the C-355 on the UK market, although the colour scheme was changed with yellow bonnet and mudguards, blue engine and chassis and cream wheels and cab.

Ursus later went on to produce smaller Massey Ferguson models under license in the early 1990s.

41

Assembled and/or
manufactured in:
Czechoslovakia
Current situation:
Zetor is still
producing tractors.

Zetor Crystal 12011

Zetor was one of the first makes from the Eastern Bloc to be imported into the UK, making their debut in the mid '60s. The first Crystal models were produced in 1968 and reached British shores four years later.

In 1973 Zetor launched the powerful Crystal 12011 in Britain. It was an advanced tractor for its time, with a particularly large and comfortable flat-deck cab. It was powered by Zetor's own 6.84-litre six-cylinder engine which produced 120 hp and drove through an 8 forward and 4 reverse transmission.

Co-operation with fellow Eastern Bloc manufacturers resulted in Ursus, who produced the back ends for the Crystal tractors, assembling almost identical tractors in Poland under their own name.

From 1981 the Crystal range was built at the ZTS Martin factory in what is now Slovakia, which is where this example was made.

Acknowledgements

I would like to thank everyone who has helped in producing this book.

Author's Note

Manufacturers' names are included for identification purposes only. All information and technical specifications are taken from the manufacturers' own sales literature, given in good faith and should only be used as a guide. The author cannot be held responsible for any errors.

Horsepower figures are given, wherever possible, as rated engine power, although other measurements may also be used.